The Metric World of SI Units

PREFACE

Express the knowledge in a manner that can be measured

The author is a mechanical-cum-nuclear engineer who has taught nuclear engineering and SI units and promoted the peaceful use of nuclear power since 1959. His six-decade journey with SI has been one of learning, un-learning, and re-learning as the system evolved. SI remains a remarkable unifying gift to humanity across all fields.

This edition includes the latest SI standards and will interest anyone involved in technical or commercial work. Born in India, the author vividly recalls the country's shift to metric units in the mid-1950s, when even his father—a small grocer with no formal schooling—enthusiastically adopted the new metric rupee of 100 paise, replacing the ancient 64-paise rupee.

The author also pursues a lifelong hobby of studying India's ancient measurement systems from the Vedic era, when sages and scientists were one. For the Indian masses, adopting the metric system proved a surprisingly simple transition.

ravirustagi16@gmail.com 2026-05-07

Table of Contents

LIST OF TABLES

1. INTRODUCTION

The book serves as a practical guide to implementing the modern International System of Units (SI), in contrast to the earlier FPS (foot–pound–second), CGS (centimeter–gram–second), and MKS (meter–kilogram–second) systems that preceded it.

Its 16 chapters are organized to give readers a quick and coherent understanding of quantities, units, dimensions, rules, and symbols. Each chapter includes worked examples and problem sets to reinforce learning and ensure clarity.

BRIEF HISTORY/ TIMELINE OF UNITS

Year	Event / Development
1799	Birth of the metric system during the French Revolution; platinum meter and kilogram standards created.
1824	British Imperial System introduced; U.S. Customary System later derived from the FPS system.
1832	Carl Friedrich Gauss proposes the CGS (centimeter–gram–second) system.
1874	British Association for the Advancement of Science recommends CGS as the standard metric system.
1880	CGS expanded with ohm, volt, and ampere to address electricity and magnetism.
1939	MKSA system introduced: meter, kilogram, second, and ampere as base units.
1960	International System of Units (SI) formally launched as the modern coherent metric system.

The inter-comparison of the traditional CGS, MKS, FPS, and USCS systems is summarized in Table 1 below. These are gravitational units based on four quantities viz Force, Length, Time, and Mass, three of these are the base units and one is a derived unit.

Table 1: Traditional Engineering Units

Name of Unit	Dimensional symbol	MKS system	Metric "absolute" or the CGS system	British FPS system	US customary or the USCS system
Force	F	**kgf***	**dyne***	**lbf***	lb
Length	L	m	cm	ft	ft
Time	T	sec	sec	sec	sec
Mass	M	kg	gm	lb	**slug***

* These are derived unit

Notes: In the MKS (meter-kilogram-second) advanced gravitational metric system, which has been used for many years in non-English-speaking countries, including India, the kilogram kg has been used both as a unit of force (kgf) as well as a unit of mass (kg). This has created confusion.

The units to measure the four fundamental quantities cannot be chosen independently, as these must be consistent with Newton's second law of motion, which states the applied force is proportional to the product of mass and acceleration, i.e.

GRAVITATIONAL UNITS

$F \propto m \cdot a$, or $F = m \cdot a/g_c$
where m = mass, a = acceleration, g_c = constant of proportionality

Table-2 on next page brings out the urgent need of change in favor of new coherent metric system, popularly known as the SI units.

Principle of coherence: In SI, all derived units are coherent. Coherent units are products of powers of one or more base units that include no numerical factor other than one.

Ex: 1 J = 1 N · 1 m

Coherence makes it possible to manipulate units with simple algebra.

change is the only constant in making progress

Table 2: Newton's Second Law of Motion- constant g_c

Gravitational units

QUANTITY (Dimensional Symbol)	**FPS Old British system**	**U.S.C.S** New Advanced FPS system	**MKS Old metric system**	**SI* New Metric System**
Length (L)	foot (ft)	foot (ft)	meter (m)	meter (m)
Mass (M)	Pound (lb)	slug (derived- no symbol)	kilogram (kg)	kilogram (kg)
Time (T)	second (sec)	second (sec)	second (sec)	second (s)
Force (F)	Pound-force lbf (derived)	lb pound	kilogram-force kgf (derived)	newton (N)
Acceleration (a) by definition	32.2 ft/sec^2	1 ft/sec^2	9.81 m/sec^2	1 m/s^2
Proportionality constant g_c (dimensionless)	32.2	1	9.81	1

A cause-effect relationship of all unit systems is presented below.			
System	**Force**	**Exerted on mass of**	**Produces an acceleration of**
USCS	1 lbf	1 slug	1 ft/sec^2
FPS	1 lbf	1 lbm	32.2 ft/sec^2
SI	**1 N**	**1 kg**	**1 m/s^2**
MKS	1 kgf	1 kgm	9.81 m/s^2
CGS	1 dyne	1 gm	1 cm/sec^2

* MKS system: 1 kgf = 1 kgm x 9.81 m/s^2 …….non-coherent

SI system: 1 N = 1 kg x 1 m/s …….coherent

2. SI SEVEN (7) BASE UNITS

In 1960, SI was formally introduced as the new metric system, called Le Systeme International d'Unites in French (abbreviated SI).

SI selected seven base units, corresponding to seven physical quantities, and natural constants.

Table 3: Seven base SI units in 1960.

Quantity	Unit Symbol	Name
time	*s*	second
length	*m*	meter
mass	*kg*	kilogram
electric current	A	ampere
thermodynamic temperature	K	kelvin
amount of substance	*mol*	mole
luminous intensity	*cd*	candela

Note: Symbols are written in lower case, except when the unit is derived from a proper name, such as A for ampere, and K for kelvin.

2019 Revision of SI Base Units

The 2019 revision of the SI was a landmark update designed to enhance the accuracy, stability, and long-term reliability of measurements. Four of the seven SI base units—the kilogram, ampere, kelvin, and mole—were redefined in terms of fundamental physical constants, a change approved by the 26th General Conference on Weights and Measures (CGPM) and implemented on May 20, 2019.

The remaining base units—the second, meter, and candela—were already linked to universal constants. This revision marked the culmination of decades of scientific research, ensuring that the SI system remains robust and adaptable to future advances in science and technology.

Table 4: 2019 Current Definitions of seven (7) SI Base Units

Unit name	Unit symbol [Dimensional symbol]	Quantity name	Definition
second	s [T]	time	The duration of 9192631770 periods of the radiation corresponding to the transition between the two hyperfine levels of the ground state of the caesium-133 atom.
meter	m [L]	length	The distance travelled by light in vacuum in 1/299792458 second
kilogram	kg [M]	mass	The kilogram is defined by setting the Planck constant *h* to 6.62607015×10^{-34} J·s (J = kg·m^{2}·s^{-2}), given the definitions of the meter and the second.
ampere	A [I]	electric current	The flow of $1/1.602176634\times10^{-19}$ times the elementary charge *e* per second, which is approximately $6.2415090744\times10^{18}$ elementary charges per second.
kelvin	K [Θ]	temperature	The kelvin is defined by setting the fixed numerical value of the Boltzmann constant *k* to 1.380649×10^{-23} J·K^{-1}, (J = kg·m^{2}·s^{-2}), given the definition of the kilogram, the metre, and the second.
mole	Mol[N]	amount of substance	The mole is the amount of substance, containing $6.022\,140\,76\times10^{23}$ elementary entities. This number is the fixed numerical value of the Avogadro constant, N_A, when expressed in the unit mol^{-1}.
candela	Cd[J]	Luminosity	The luminous intensity, in a given direction, of a source that emits monochromatic radiation of frequency 5.4×10^{14} hertz and that has a radiant intensity in that direction of 1/683 watt per steradian.

3. SI BASE AND DERIVED UNITS

The derivations of derived units from base units are explained below.

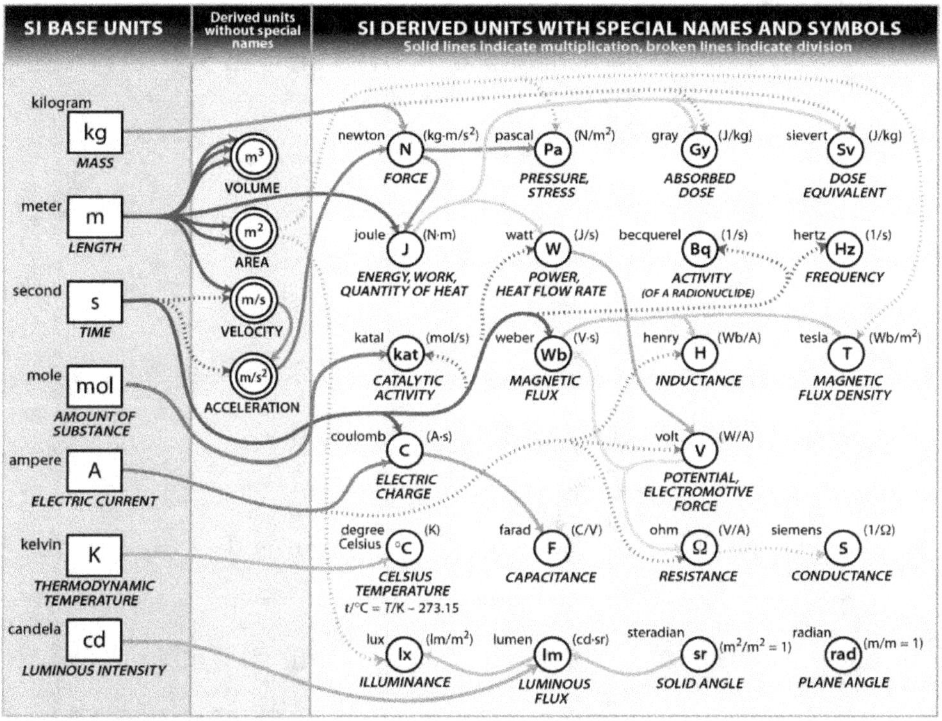

Note: In the above, 1 Pa = 1 N/1 m^2.........coherent

4. SI DERIVED UNITS

Derived units are defined as products of powers of the base units, when the numerical factor of this product is one. Examples of typical derived units are given in Table-5.

Table 5: Dimensions of derived units in SI

Dimension	**SI unit**	**SI base unit**
Speed [L/T]	meter/sec	[m/s]
Frequency [1/T]	hertz [Hz]	[1/s]
Thrust* [$M \cdot L \cdot T^{-2}$]	newton [N]	[$kg \cdot m \cdot s^{-2}$]
Energy [$M \cdot L^{2} \cdot T^{-2}$]	joule [J]	[$N \cdot m = kg \cdot m^{2} \cdot s^{-2}$]
Power [$M \cdot L^{2} \cdot T^{-3}$]	watt [W]	[$J/s = kg \cdot m^{2} \cdot s^{-3}$]
Electric Charge [$I \cdot T$]	coulomb [C]	[$A \cdot s$]
Voltage [$M \cdot L^{2} \cdot T^{-3} \cdot I^{-1}$]	volt [V]	[$J/C = kg \cdot m^{2} \cdot s^{-3} \cdot A^{-1}$]

Table 6: Examples of SI derived units expressed in terms of seven base units		
Derived quantity	**Name, typical symbol**	**Symbol**
area	square meter, A	m^2
volume	cubic meter, V	m^3
speed, velocity	meter per second, v	m/s
acceleration	meter per second squared, a	m/s^2
wave number	reciprocal meter, σ	m^{-1}
surface density	kilogram per square meter, ϱ_A	kg/m^2
specific volume	cubic meter per kilogram, v	m^3/kg
current density	ampere per square meter, j	A/m^2
magnetic field strength	ampere per meter, H	A/m
amount-of-substance concentration	mole per cubic meter, c	mol/m^3
luminance	candela per square meter, L_v	cd/m^2
mass concentration	kilogram per cubic meter, ϱ, γ	kg/m^3

***Principle of coherence:** Coherent units are products of powers of one or more base units that include no numerical factor other than one. Ex: 1 J = 1 N · 1 m

Coherence makes it possible to manipulate units with simple algebra. Table below includes 22 derived units in common use.

Table 7: The 22 derived SI units with special names and symbols

Quantity	Name	SI Symbol		SI Unit
plane angle	radian	rad	-	$m \cdot m^{-1} = 1$
solid angle	steradian	sr	-	$m^2 \cdot m^{-2} = 1$
frequency	hertz	Hz	-	s^{-1}
force	newton	N	-	$m \cdot kg \cdot s^{-2}$
pressure, stress	pascal	Pa	N/m^2	$m^{-1} \cdot kg \cdot s^{-2}$
energy, work, quantity of heat	joule	J	$N \cdot m$	$m^2 \cdot kg \cdot s^{-2}$
power, radiant flux	watt	W	J/s	$m^2 \cdot kg \cdot s^{-3}$
electric charge, quantity of electricity	coulomb	C	-	$A \cdot s$
electric potential difference, emf	volt	V	W/A	$m^2 \cdot kg \cdot s^{-3} \cdot A^{-1}$
capacitance	farad	F	C/V	$m^{-2} \cdot kg^{-1} \cdot s^4 \cdot A^2$
electric resistance	ohm	Ω	V/A	$m^2 \cdot kg \cdot s^{-3} \cdot A^{-2}$
electric conductance	siemens	S	A/V	$m^{-2} \cdot kg^{-1} \cdot s^3 \cdot A^2$
magnetic flux	weber	Wb	$V \cdot s$	$m^2 \cdot kg \cdot s^{-2} \cdot A^{-1}$

Table-7 con't				
Quantity	**Name**	**Symbol**		**SI UnitT**
magnetic field strength	tesla	T	Wb/m^2	kg·s^{-2}·A^{-1}
inductance	henry	H	Wb/A	m^2·kg·s^{-2}·A^{-2}
temperature	celsius	°C	-	K
catalytic activity	katal (mole per second)	kat	-	mol· s^{-1}
luminous flux	lumen	lm	cd·sr	m^2·m^{-2}·cd = cd
illuminance	lux	lx	lm/m^2	m^2·m^{-4}·cd = m^{-2}·cd
activity (of a radionuclide)	becquerel	Bq	-	s^{-1}
absorbed dose, specific energy (imparted), kerma	gray	Gy	J/kg	m^2·s^{-2}
dose equivalent	sievert	Sv	J/kg	m^2·s^{-2}

NOTE: The radian and steradian are strictly dimensionless units of plane and solid angle respectively and may be used advantageously in expressions for derived units.

In practice, the symbols *rad* and *sr* are used where appropriate for angular measure. Symbol *Rad* with capital R is used in nuclear radiation measurement. In photometry, the name steradian and the unit symbol *sr* are usually retained in expressions for derived units. Other quantities expressed in sieverts are ambient dose equivalent, directional dose equivalent, personal dose equivalent, and organ equivalent dose in radiation measurement.

5. NON-SI UNITS: ACCEPTED FOR USE WITH THE SI

It is recognized that some non-SI units are still used and are expected to be used for many years. Therefore, the **CIPM** (International Committee for Weights & Measures) has accepted some non-SI units for use with the SI; these are listed in Table 8. The information is presented in respect of their symbol and preferred unit.

Table 8. Non-SI units accepted in SI

Quantity	Name of unit	Symbol for unit	Value in SI units
time	minute	min	1 min = 60 s
	hour	h	1 h = 60 min = 3 600 s
	day	d	1 d = 24 h = 86 400 s
Length, plane and phase angle	astronomical unit	au	1 au = 149 597 870 700 m
	degree	$^{\circ}$	$1^{\circ} = (\pi/180)$ rad
	minute	'	1' = (1/60°) = (π/ 10 800) rad
	second	"	1" =(1/60') =(π/ 648 000) rad
area	hectare	ha	1 ha = 1 hm^2 = 10^4 m^2
volume	liter	L	1 L = 1 dm^3 = 10^3 cm^3 = 10^{-3} m^3
mass	metric ton, tonne	t	1 t = 10^3 kg
	dalton (atomic mass unit)	Da	1Da=1.6605 $\times 10^{-27}$ kg
energy	electronvolt	eV	1 eV = 1.602 x 10^{-19} J
logarithmic ratio quantities	neper	N_p	In Physics, N_p and B are used to express a ratio of powers, currents or noise levels.
	bel	B	
	decibel	dB	

SI has truly demonstrated the distinction between competition, and selfless co-operation

5. NON-SI UNITS (continued...)

NIST: The **National Institute of Standards and Technology** (**NIST**) is an agency of the United States Department of Commerce whose mission is to promote American innovation and industrial competitiveness.

NIST activities are organized into physical science laboratory programs that include nanoscale science and technology, engineering, information technology, neutron research, space research, material measurement, and physical measurement.

For example, other units are not strictly SI, but NIST has accepted for use with the SI, units listed in Table 8 with the SI except for the curie, roentgen, rad, and rem.

Because of the continued wide use of these units in the United States, NIST still accepts their use with the SI.

Electromagnetic units (EMU): what is an electrical charge?
These quantities have always been in metric units. SI unit of electrical charge is the coulomb C, defined as the product A·s.
Electromagnetic force (in newtons) between point charges Q_1 and Q_2 is given by Coulomb's Law

$Fe = \frac{kQ1Q2}{r^\wedge 2}$ (Coulomb's law)

<u>SI is truly universal</u>
SI units with prefixes and rules for writing are quickly boosting the growth in contemporary areas of science and technology worldwide such as space research, nuclear and environmental science, medical and pharmaceutical industry, and more. This is mainly because SI measurements are consistent, reliable and verifiable; hence fast becoming a universal language of measurement in all branches of knowledge.

6. SI PREFIXES

It is now established that SI prefixes are extremely important in the study of science as well as commerce. The popular 20 prefixes are short letter symbols or names attached in front of the unit's name further simplifies SI.
These are listed in the Table 9 below. The advantage of a prefix is to eliminate the confusing awkward and hard to pronounce unit names.

Table 9. SI prefixes

Factor	Name	Symbol	Factor	Name	Symbol
10^{24}	yotta	Y	10^{-1}	deci	d
10^{21}	zetta	Z	10^{-2}	centi	c
10^{18}	exa	E	10^{-3}	milli	m
10^{15}	peta	P	10^{-6}	micro	µ
10^{12}	tera	T	10^{-9}	nano	n
10^{9}	giga	G	10^{-12}	pico	p
10^{6}	mega	M	10^{-15}	femto	f
10^{3}	kilo	k	10^{-18}	atto	a
10^{2}	hecto	h	10^{-21}	zepto	z
10^{1}	deka	da	10^{-24}	yocto	y

Note 1: Normally, it is enough to use from tera symbol T (10^{12}) to nano symbol n (10^{-9}). For example,
Diameter of the Sun = 1.4 Gm = 1.4 x 10^9 m
Diameter of the Earth = 12.7 Mm = 12.7 x 10^6 m
To compare the two diameters, we divide the larger by the smaller.
(1.4 x 10^9)/(12.7 x 10^6) = 110 the answer is rounded to two significant digits, because the earth is not a perfect circle, and the diameters vary.
To solve by a calculator, we enter: 1.4 EXP 9 ÷ 12.7 EXP 6 = **110.2**

In the ancient Hindu books of India, the rounded number is 108.

Note 2: A SI based unit may be combined with any of the prefixes to create a unit that is most effective for what is being measured or expressed. For example, you would not measure the distance from NYC to LA in meters. Instead, you would use 4139 kilometers (km).

Distance from earth to moon is better expressed as megameters 363 Mm.

Note 3: In commerce, the 2025 union budget of the Government of India, total expenditure is projected at rupees 50.65 x 10^{12} rupees = 50.65 lakh crores. The Indian public better comprehends 50.65 lakh crores, instead of 50.65 Terra rupees. It is the SI spirit of prefixes that is retained.

7. SIGNIFICANT DIGITS

Definition: The number of *reliably* known digits, as measured in any physical quantity is called the number of significant digits, and this determines the accuracy of the number. See the Table below.

Table 10. Significant digits

Number as written	Number of significant digits	Implied range (accuracy)
20 (estimate)	1	20 being a rough estimate, 0 is a place holder only.
20	2	20 ± 1, or 19 – 21
20.0	3	20 ± 0.1, or 19.9 – 20.1
0.00241*	3	0.00240 – 0.00242
24100**	May be 3,4 or 5	See rule [3)]
241×10^6	3	$240 \times 10^6 - 242 \times 10^6$
2.41×10^{-3}	3	$2.40 \times 10^{-3} - 2.42 \times 10^{-3}$

Rules: There are five (5) rules governing significant digits:

1. The significant digits in a number include the left-most non-zero digits to the right-most digit written.
2. If the zeros occur at the beginning of a number that is less than one, then the zeros are not significant. i.e. one zero before the decimal in number 0.00241 is merely a placeholder to show where the decimal point is placed, but these zeros do not count as significant figures.
3. **Engineering notation:** ** 24100 may have three (241), four (2410) or five (24100) significant digits. To avoid these ambiguities, engineering notation is used to report the result in multiples of (10^3). This means if 24100 has five significant digits, it is written as 24.100(10^3).

4. If 24100 has three significant digits, it is written as 24.1(10^3). Likewise, 0.000241 having three significant digits, can be written as 0.241(10^{-3}), or as 241(10^{-6}).
5. **Rounding off**: The numbers should be rounded off to the desired number of significant digits, before expressing in the engineering notation. As a general rule, if the numerical figure ends in 5 or greater it is rounded up, and a number less than 5 is rounded down. Suppose the number 2.407 is rounded off to three significant digits. Since the last digit (7) is >5, the third number (0) is rounded up to 2.41. Now if we round off 2.41 to two significant digits, the third digit (1) is < 5, we will get the answer 2.4. There is a special case for any number having 5 with zeros following it: if the digit preceding the 5 is an even number, then this digit is not rounded up, but if the digit preceding the 5 is an odd number, then this digit is rounded up. For example, 45.25 rounded off to three significant digits becomes 45.2 and not 45.3. However, 45.35 becomes 45.4 and 0.04535 becomes 0.045.
6. **Significant digits/figures** in the final result should be the same as the least significant input value.

For example,
if we calculate the area of a rectangle 10.4 x *5.3. The result of multiplication 55.12 should be rounded off to 55 to two significant figures. Similarly, 2.5* x *1.6 = 4 must be expressed as 4.0*
- Another case: sin 60° means two significant figures. Therefore, the calculator result: 0.866025403 must be rounded off as 0.87.
- Keep one or more extra significant figures throughout the calculation, and round off only in the final result.

In the above example, if we initially round off 10.4 as 10 to two significant figures, the result of 10 x 5.3 multiplication will be 53 instead of 55.12. This shows rounding off be done in the final answer.

Note: In engineering mechanics, the loads and geometrical data can be conveniently measured to three significant digits, therefore the sequence of calculations must be performed to four or more significant digits, and the final result rounded off to three significant digits.

To convert 30.0 x 10^6 psi to SI:

You should recall this is the modulus of elasticity E value for mild steel, and is obtained in a tensile test, where stress σ is rounded off to 3 significant figures, and strain ε to 4 significant figures.

Therefore, E must be expressed in 3 significant figures, as 30.0 x 10^6.

The calculated SI result 206.85 GPa needs also to round off to 207 GPa.

- **Accuracy** refers to how close a measurement is to the true or reference value.
- how repeatable a measurement is?

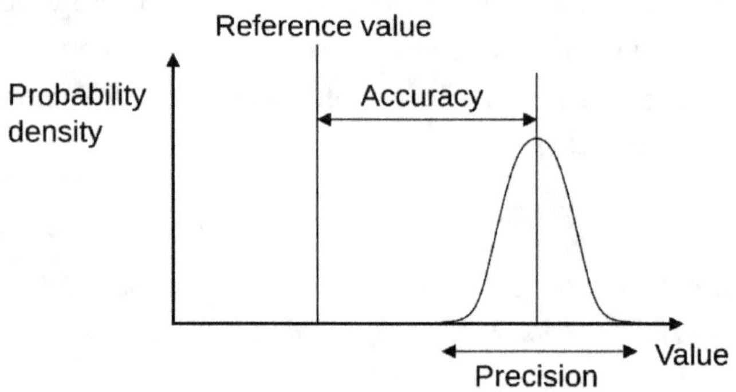

8. SI RULES AND STYLE CONVENTIONS

Symbols and prefixes are the same in all languages. Symbols are written in lowercase, except when the unit is derived from a proper name.

Table 11. SI units, styles, and conventions

#1 **General**	Only units of the SI and those units recognized for use with the SI are used to express the values of quantities. Equivalent values in other units are given in parentheses following values in acceptable units only when deemed necessary for the intended audience.
#2 **Abbreviations**	Abbreviations such as sec, cc, or mps are avoided and only standard unit symbols, prefix symbols, unit names, and prefix names are used. **proper:** s or second; cm^3 or cubic centimeter; m/s or meter per second Improper sec; cc; mps
#3 **Plurals**	Unit symbols are unaltered in the plural. **proper:** l = 75 cm improper: l = 75 cms
#4 **Punctuation**	Unit symbols are not followed by a period unless at the end of a sentence. **proper:** The bar is 75 cm long. improper: The bar is 75 cm. long.

#5 Multiplication & division

A space or half-high dot is used to signify the multiplication of units. A solidus (*i.e.,* slash), horizontal line, or negative exponent is used to signify the division of units. The solidus must not be repeated on the same line unless parentheses are used.

proper:

speed of sound is 344 m/s (meters per/second)
the decay rate of ^{113}Cs is 21 ms^{-1},
m/s, m·s^{-2}, m·kg/(s^{3}·A),
m·kg·s^{-3}·A^{-1}
m/s, m s^{-2}, m kg(s^{3} A), m kg s^{-3} A^{-1}

improper:

speed of sound is 344 ms^{-1} (reciprocal milliseconds)
The decay rate of ^{113}Cs is about 21 m·s^{-1} (meters per second)
m ÷ s, m/s/s, m·kg/s^{3}/A

#6 Typeface

Variables and quantity symbols are in italic type. Unit symbols are in roman type. Numbers should generally be written in roman type. These rules apply irrespective of the typeface used in the surrounding text.

proper:

She exclaimed, "***That dog weighs*** 10 kg***!***"
t = 3 s, where *t* is time and s is second
T = 22 K, where *T* is thermodynamic temperature, and K is kelvin

improper:

She exclaimed, "***That dog weighs 10 kg***"
t = 3 s, where t is time and s is second
T = 22 K, where T is thermodynamic temperature, and K is kelvin

#7 **Typeface**

Superscripts and subscripts are in italic type if they represent variables, quantities, or running numbers. They are in roman type if they are descriptive.

subscript category	**typeface**	**proper usage**
quantity	*italic*	c_p, specific heat capacity at constant pressure
descriptive	roman	m_p, mass of a proton
running number	*italic*	$x = \bar{X} = \frac{1}{n}\sum_{i=1}^{n} X_i$

#8 **Abbreviations**

The combinations of letters "ppm," "ppb," and "ppt," and the terms part per million, part per billion, and part per trillion, and the like, are not used to express the values of quantities.

proper: 2.0 µL/L; 2.0 x 10^{-6} V;
4.3 nm/m; 4.3 x 10^{-9} l;
7 ps/s; 7 x 10^{-12} t,
where V, l, and t are the quantity symbols for volume, length, and time.

improper: "ppm," "ppb," and "ppt," and the terms part per million, part per billion, and part per trillion, and the like

#9 Modifications

Unit symbols (or names) are not modified by the addition of subscripts or other information. The following forms, for example, are used instead.

proper: $V_{max} = 1000$ V
a mass fraction of 10 %

improper: $V = 1000$ V_{max}
10 % (*m*/*m*) or 10 % (by weight)

#10 % (percent)

The symbol % is used to represent simply the number 0.01.

proper: $l_1 = l_2(1 + 0.2\ \%)$, or
$D = 0.2\ \%$,
where D is defined by the relation $D = (l_1 - l_2)/l_2$.

improper: the length l_1 exceeds the length l_2 by 0.2 %

#11 Information

Information is not mixed with unit symbols or names.

proper: the water content is 20 mL/kg

improper: 20 mL H_2O/kg
20 mL of water/kg

#12 **Notation**	It is clear which mathematical operation applies to the value of a quantity. **proper:** 35 cm x 48 cm 1 MHz to 10 MHz or (1 to 10) MHz 20°C to 30°C or (20 to 30)°C 123 g ± 2 g or (123 ± 2) g 70 % ± 5 % or (70 ± 5) % 240 ± 10 % V improper: 35 x 48 cm 1 MHz-10 MHz or 1 to 10 MHz 20 °C- 30 °C or 20 to 30 °C With (Celsius) temperature, there is no space. e.g. 20°C is proper (not: 20 °C)
#13 **Symbols & Names**	Unit symbols and unit names are not mixed and mathematical operations are not applied to unit names. **proper:** kg/m^3, $kg \cdot m^{-3}$, or kilogram per cubic meter improper: kilogram/m^3, kg/cubic meter, kilogram/cubic meter, kg per m^3, or kilogram per $meter^3$.
#14 **Numerals & Unit Symbols**	Values of quantities are expressed in acceptable units using Arabic numerals and symbols for units. **proper:** m = 5 kg the current was 15 A

improper: m = five kilograms
m = five kg
the current was 15 amperes

#15 Spacing

There is a space between the numerical value and unit symbol, even when the value is used in an adjectival sense, except in the case of superscript units for plane angle.

proper: a 25 kg sphere
an angle of 2° 3′ 4″
If the spelled-out name of a unit is used, the normal rules of English apply: "a roll of 35-millimeter film."

improper: a 25-kg sphere
an angle of 2 ° 3 ′ 4 ″

#16 Spacing

The digits of numerical values having more than four digits on either side of the decimal marker are separated into groups of three using a thin, fixed space counting from both the left and right of the decimal marker. Commas are not used to separate digits into groups of three.

proper: 15 739.012 53

improper: 15739.01253
15,739.012 53

#17 Quantity equations

Equations between quantities are used in preference to equations between numerical values, and symbols representing numerical values are different from symbols representing the corresponding quantities.

When a numerical-value equation is used, it is properly written and the corresponding quantity equation is given where possible.

proper: $(l/\text{m}) = 3.6\ [v/(\text{km/h})](t/\text{s})$

improper: $l = 3.6\ vt$, accompanied by text saying, "where l is in meters, v is in kilometers per hour, and t is in seconds"

#18
Standard symbols

Standardized quantity symbols are used. Similarly, standardized mathematical signs and symbols are used. More specifically, the base of "log" in equations is specified when required by writing $\log_a x$ (meaning log to the base *of* x), lb x (meaning $\log_2 x$), ln x (meaning $\log_e x$) , or lg x (meaning $\log_{10} x$).

proper: tan x
R for resistance
A_r for relative atomic mass

improper: tg x for tangent of x
words, acronyms, or ad hoc groups of letters

#19
Weight vs. mass

When the word "weight" is used, the intended meaning is clear. (In science and technology, weight is a force, for which the SI unit is the newton; in commerce and everyday use, weight is usually a synonym for mass, for which the SI unit is the kilogram.)

#20 **Quotient quantity**	A quotient quantity is written explicitly. **proper:** mass divided by volume improper: mass per unit volume
#21 **Object & quantity**	An object and any quantity describing the object are distinguished. (Note the difference between "surface" and "area," "body" and "mass," "resistor" and "resistance," "coil" and "inductance.") **proper:** A body of mass 5 g improper: A mass of 5 g

Note: Use spaces instead of commas to put digits into easily readable blocks of three. e.g. 3 407 359.38 (not: 3,407,359.38). However, this does not apply to years, telephone numbers, street numbers, etc.
e.g. 1985 (not: 1 985); Tel. 205-8569 (not: 205-8 569).

#22 **Obsolete Terms** The obsolete terms normality, molarity, and molal and their symbols N, M, and m are not used.

proper: amount-of-substance concentration of B (more commonly called concentration of B), and its symbol c_B and SI unit mol/m^3 (or a related acceptable unit)
molality of solute B, and its symbol b_B or m_B and SI unit mol/kg (or a related unit of the SI)

improper: normality and the symbol *N*, molarity and the symbol M
molal and the symbol m

#23 **When writing a date**

1. **American format**: month-day-year (May 1, 2022), with a comma between day and year and numeric only 05-01-2022
2. **British format**: day-month-year (1 May 2022), with no comma between month and year and numeric only 01-05-2022
3. **International format: YYYY-MM-DD (2022-05-01)**
   YYYY year in the usual Gregorian calendar (2022), MM month of the year (05), DD day of the month (01)
4. You may also show the **day of the week** (Sunday May 1, 2022).
5. **Consistency** is important when using slashes, hyphens, or dots.

#24 **When writing a time**

1. When you write time, **use numbers and be specific about the part of the day**. For example, 8:00 AM for morning times and 3:00 PM for afternoon or evening times.
2. If you're mentioning a time on the hour, it's correct to write it as 4:00, not just 4. For minutes after the hour, write them as they are: 2:30 PM.
3. For instance, all the following would be acceptable when time is part of a sentence-

*She gets up at **six** in the morning every day.*
*She gets up at **6:00** in the morning every day.*
*She gets up at **6 o'clock** in the morning every day.*

4. As such, we use AM and PM to indicate the time of day when we are using a 12-hour clock. We can see how this works more clearly if we write out various times of day in both 12-hour and 24-hour formats-

5. Time Examples

Time	12-Hour Clock	24-Hour Clock
Midnight	12:00 AM	00:00
Half past eight in the morning	08:30 AM	08:30
Midday	12:00 PM	12:00
Three in the afternoon	03:00 PM	15:00
Eight Thirty in the evening	08:30 PM	20:30

9. CONVERSION FACTORS

Grouped as Table 12 and 13, handy conversion factors from the traditional units to the SI units are provided in this chapter.
To warm up, the chapter is sub-divided into two parts

(i) Basic units of length, mass, and time
(ii) Derived units in the alphabetic order

Whereas the conversion factors are given in tabular form, the yellow shaded column represents the units in the SI system rounded off to not more than four significant digits.

Table 12: Conversion Factors for common base units
LENGTH, MASS, TIME

LENGTH:

cm	**m (SI)**	ft
1	**0.01**	0.0328
100	**1**	3.28
30.48	**0.305**	1

Useful Notes:

1 inch = 2.54 cm = 0.0254 m
1 micron = 0.001 mm
1 mil = 0.001 in.
1 Å = 0.1 nm (A= angstrom unit)
try to **avoid** such units and names

MASS:

t long ton	**kg (SI)**	lb	slug
1	**1000**	2205	68.54
0.001	**1**	2.205	0.0685
0.000454	**0.454**	1	0.0031
0.0146	**14.59**	32.2	1

In MKS gravitational system, the kilogram has been used both as a unit of force and as a unit of mass.

TIME: Second (symbol s) is the SI unit; the following units of time are used for convenience.

1 min = 60 s
1 h = 3600 s
1 d = 86.4 ks … kilo seconds
1 a = 31.53 Ms…Mega seconds

Table 13: CONVERSION FACTORS FOR COMMON DERIVED UNITS IN ALPHABETIC ORDER

ACCELERATION:

1 ft/s^2 = 0.305 m/s^2

AREA:

cm^2	**m^2 (SI)**	in^2	ft^2
1	**10^{-4}**	0.155	1.076 x 10^{-3}
10^4	**1**	1.55 x 10^3	10.76
6.452	**6.45 x 10^{-4}**	1	6.942 x 10^{-3}
929.36	**0.093**	144	1

Useful notes:

- Area is the quantity of surface, and its SI unit is square meter (m^2). Proper way is that the word 'square' is pronounced first. For example, 3 m^2 is pronounced as 3 square meters, not $(3m)^2$, as this would mean 9 square meters.
- A hectare (symbol ha) is a unit of area equal to **10,000 square meters**. To determine the size of a football field in hectares, one needs to calculate the field's area in square meters and then divide by 10,000.
- 1 acre = 4840 sq yd = 4051.1 sq m = 0.405 ha

The acre is a unit of land area used in the British imperial and the United States customary systems.

NUCLEAR X-SECTION barn (symbol **b**)**:**

1 b = 10^{-28} m^2

BURN-UP (NUCLEAR FUEL):

MW·d/t	**GJ/kg (SI)**	MW·h/kg
1	**0.086**	0.024
11.57	**1**	0.277
41.67	**3.61**	1

CONCENTRATION:

g/L	**kg/m^3 (SI)**	ppm*
1	**1**	1000
0.001	**0.001**	1
0.01425	**0.01425**	14.25

Strictly speaking, kg/m^3 is a measure of density. Remember that the expressions g/cm^3, kg/m^3, and Mg/m^3 are equivalent.
*ppm (parts per million) relates to mixtures, solutes and solvents. Such quantities are traditionally expressed in parts per cent ($\% = 10^{-2}$), or per thousand ($‰ = 10^{-3}$), parts per million ($ppm = 10^{-6}$), parts per billion ($ppb = 10^{-9}$). For example, seawater averages 35 g of salt/kg of water (35‰)
Problem: Express 50 ppm concentration in SI unit (mass per mass)
Solution: 50 mg/kg

CORROSION RATE:

Corrosion rates in the U.S. are normally calculated using **mils per year** (mil=one thousandth of inch). To convert the corrosion rate between the mils per year (mil/a) and the metric equivalent millimeter per year (mm/a), use the following equation
1 mil/a = 0.0254 mm/a = 25.4 µm/a

In other words, the corrosion rate is based on the number of mills (thousandths of an inch) penetrated each year. To calculate the corrosion rate, you'll need the following information:

1. **Weight loss:** The decrease in metal weight during a reference time period. mg or kg. Or loss in volume expressed as thickness mil.
2. **Density:** The density of metal ϱ.
3. **Area:** The total initial surface area of the metal piece.
4. **Time:** The length of reference time period- second(s), day(d) or year(a).

CORROSION RATE con't:

mil/a	kg/(m² · s) (SI)	mg/(dm² · d)	in/a
1.44 / ϱ	**1.157 x 10^{-9}**	1	0.00144/ ϱ
1000	**(8.054 x 10^{-7})** ϱ	695.89 ϱ	1
1.24 x10^{9}/ ϱ	**1**	8.64 x 10^{8}	1.24 x 10^{6} / ϱ

DENSITY: density = (mass ÷ volume)

g/ cm³	kg/ m³ (SI)	lb_m / ft³
1	**10^3**	62.29
10^{-3}	**1**	0.0622
0.0161	**16.08**	1

Problem: What is the density of earth, expressed as Mg/m³?
Data: Mass of earth = 5.98 · 10^{24} kg, radius of earth = 6.37 Mm
Solution: density ϱ = mass/volume
mass = 5.98 · 10^{24} kg = 5.98 · 10^{21} Mg
volume = 4/3·π· $(6.37·10^6)^3$ = 1.0826 · 10^{21} m³
density of earth = 5.98/1.0826 = **5.5 Mg/m³** (2 significant digits)
density of Pyramid of Giza = **2.35 Mg/m³** **ref. p 40**
Conclusion: The core of earth is mostly metallic.
In comparison, density of human body = **1.01 Mg/m³**

ENERGY:

kcal	kJ (SI)	BTU
1	**4.187**	3.968
0.239	**1**	0.948
0.252	**1.054**	1

ENTHALPY: Same as Internal energy

useful notes on energy: 1 erg = 10^{-7} J, 1 BTU = 778 ft lb

1 J = 1 W·s , 1 eV = 160.2 zJ…..(zetta-Joule)

Nutrition label, and exercise: Calorie or k-calorie represent the same amount of energy. Calories may be expressed as kilojoules (kJ).
One calorie, or kilocalorie are equal, and = 4.184 kJ.
For more information, refer to Appendix- 3.4, p.50

FLOW RATE:

flow rate (mass):

lb min^{-1}	**kg/s (SI)**	**kg/h**
0.036	**2.77×10^{-4}**	1
132	**1**	3600
1	**7.575×10^{-3}**	27.27

useful notes:

The gallon is a unit to measure volume of liquids in British Imperial unit and the US customary units, 1 Imp. gallon = 1.201 US gallon

1 lb/hr flow of water = 2.00×10^{-3} USGPM (water density =1)

1 USGPM = 3.785 L/min; 1 IGPM = 4.546 L/min…liquid flow

US liquid gallon = 231 cub.in; US dry gallon = 268.8 cub. in.

flow rate (volumetric):

L/sec	**m^3 /s (SI)**	**ft^3 /hr**	**L/min**
1	**0.0010**	1.27×10^2	60
10^3	**1**	35.31	6.04×10^4
7.86×10^{-3}	**7.86×10^{-6}**	1	4.72×10^{-1}
16.6×10^{-2}	**1.67×10^{-5}**	2.12	1

useful notes:1 cfm = 2.78×10^4 cm^3/min 1 L/min = 0.035 scfm

FORCE:

kgf	N (SI)	lbf	dyne
1	**9.807**	2.205	9.807×10^5
0.102	**1**	0.225	10^5
0.4536	**4.448**	1	4.448×10^5

HEAT FLOW PER UNIT VOLUME:

kcal/(m^3 · h)	W/m^3 (SI)	BTU/(ft^3 · h)
1	**1.163**	0.034
0.86	**1**	0.029
29.24	**34.013**	1

HEAT FLUX: (same as heat flow per unit area)

kcal/(m^2 · h)	W/ m^2 (SI)	BTU/(ft^2 · h)
1	**1.163**	0.369
0.86	**1**	0.317
2.712	**3.154**	1

HEAT TRANSFER CO-EFFICIENT:

kcal/(m^2 · °C · h)	W/(m^2 · K) (SI)	BTU/(ft^2 · h · °F)
1	**1.163**	0.205
0.860	**1**	0.176
4.882	**5.678**	1

INTERNAL ENERGY: same as enthalpy

kcal/kg	**kJ/ kg (SI)**	BTU/ lb
1	**4.187**	1.8
0.239	**1**	0.43
0.555	**2.326**	1

MASS FLUX:

lb/(ft^2 · h)	**kg/(m^2 · h) (SI)**
0.2046	**1**
1	**4.8868**

MASS FLOW (Leak rate):

atm·cm^3/min	torr·L/sec	mbar·L/sec *	**Pa·m^3/sec (SI)**
1	4.56 x 10^1	60.6	**6.06**
7.90 x 10^1	1	1.33	**1.33 x 10^1**
5.92 x 10^1	7.50 x 10^1	1	**1.01 x 10^1**
592	7.5	10	**1**

* *What is leak rate? It represents how quickly gas escapes from an enclosed or evacuated volume. If the gas pressure in a 1-liter volume drops by 1 millibar within 1 second, the leakage rate is 1 mbar· L/ sec*

MOLE: The **mole** (symbol **mol**) is a unit of measurement for *amount of substance.* One mole contains exactly 6.022 14076 $\times 10^{23}$ elementary entities, which is the Avogadro number. 6.022 x 10^{23} are the number of atoms in 12 grams of carbon-12 or number of molecules in 18 grams of water.

(MASS) MOMENT OF INERTIA:

$g \cdot cm^2$	**$kg \cdot m^2$ (SI)**	$lb \cdot ft^2$
1	**10^{-7}**	2.37 x 10^{-6}
10^7	**1**	23.73
4.213 x 10^5	**0.042**	1

POWER:

kcal/h	**W (SI)**	BTU/h
1	**1.163**	3.968
0.86	**1**	3.413
0.252	**0.293**	1

notes:

1 HP = 550 ft lb/ s = $2.546 \cdot 10^3$ BTU/ h = 746 W

1 kJ = 0.239 kcal 1 kcal = 4.18 kJ

PRESSURE:

kg/cm^2	**Pa (SI)**	lb/in^2
1	**0.09807 x 10^6**	14.22
10.197 x 10^{-6}	**1**	144.95 x 10^{-6}
0.070	**6.894 x 10^3**	1

notes:
1 bar = 10^5 Pa = 10^2 kPa = 1.02 kg/cm^2 = 14.495 psi
1 psi = 6.894 kPa 1 mmHg = 1 Torr =133.32 N/m^2
1 atmosphere = 1.034 kg/cm^2 = 14.696 lb/in^2 = 1.013 bar = 33.8 ft H_2O
bar, psi, kg/cm^2 are not permitted within SI units

SPECIFIC ENTROPY: Same as SPECIFIC HEAT
SPECIFIC ENTHALPY: Same as INTERNAL ENERGY

SPECIFIC HEAT:

kcal/(kg x °C)	**kJ/(kg · K) (SI)**	BTU/(lb · °F)
1	**4.187**	1
0.239	**1**	0.239

SPEED:
Useful notes: 1 ft/ s = 0.305 m/ s 1 km/h = 0.278 m/s
(kmph is improper)
1 mi/h = 1.61 km/h = 0.448 m/s 30 mi/h = 44 ft/s
1 knot = 1 nautical mile per hour = (1852/3600) m/s = 0.5144 m/s

STRESS:

kgf/cm^2	**N/m^2 (SI)**	**lbf/in^2**
1	**0.981 x 10^5**	14.22
1.02 x 10^{-5}	**1**	14.50 x 10^{-5}
0.070	**0.069 x 10^5**	1

SURFACE TENSION:

dyn/ cm	**N/m (SI)**	**kgf/m**	**lbf/ft**
1	**10^{-3}**	1.019 x 10^{-4}	6.85 x 10^{-5}
10^3	**1**	0.102	0.069
9.81 x 10^3	**9.81**	1	0.672
14.60 x 10^3	**14.60**	1.487	1

TEMPERATURE:

The formula to convert Fahrenheit (°F) to Celsius (°C) is:
°C = (°F−32) · (5/9) In writing 0°C, leave no space between 0 and °C

0°C =273.16 K 0 °F = 459.67 °R 1 K = 1.8 °R
Zero on Kelvin or Rankine scale is absolute zero

THERMAL CONDUCTIVITY:

kcal/(h · m · °C)	**W/(m · K) (SI)**	**Btu/(h · ft · °F)**
1	**1.163**	0.672
0.86	**1**	0.578
1.488	**1.731**	1

THERMAL DIFFUSIVITY:

cm^2 /s	**m^2/ s (SI)**	**ft^2/ s**
1	10^{-4}	1.076 x 10^{-3}
10^4	1	10.76
929.37	0.093	1

THERMAL RESISTIVITY:

°C x h/ kcal	**K/ W (SI)**	**(°F · h)/ Btu**
1	0.860	0.454
1.163	1	0.528
2.205	1.895	1

TORQUE: a rotational force is called torque, symbol Letter **tau (τ)**

dyn· cm	**N · m (SI)**	**lbf · in**
1	1 x 10^{-7}	8.85 x 10^{-7}
10^7	1	8.852
1.129 x 10^6	0.113	1

VELOCITY: (Same as speed)

VISCOSITY:

Viscosity is a measure of the friction between the molecules of fluid. The **unit of viscosity** depends on whether it is absolute (dynamic) viscosity or kinematic viscosity.

Dynamic viscosity is the ratio of shear stress to shear rate in a fluid. The **SI unit of dynamic viscosity is N· s· m^{-2}** = Pa·s (pascal second).

CGS unit of dynamic viscosity is **poise**, symbol P. (1 P = 10 Pa·s)

Kinematic viscosity is the ratio of dynamic viscosity to density of the fluid, its SI unit is **m^2/s** and derived as below.

Kinematic viscosity = (dynamic viscosity)/density = kg m^{-1} s^{-1} /(kg/m^3)= m^2/s (SI Unit)

DYNAMIC VISCOSITY:

(kgf · s) / m^2	(N· s)/ m^2 (SI)	(lb· s)/ ft^2
1	**9.807**	0.206
0.102	**1**	0.021
4.86	**47.7**	1

Useful note:

1 centipoise = 10^{-2} P = 0.1 Pa·s

KINEMATIC VISCOSITY:

cm^2 /s (stokes st)	m^2 /s (SI)	ft^2 /s
1	**10^{-4}**	1.076 x 10^{-1}
930	**0.093**	1

VOLUME:

cm^3	m^3 (SI)	ft^3
1	10^{-6}	3.53×10^{-5}
10^6	1	35.31
2.83×10^4	0.0283	1

Useful Notes on volume:

8 pints = 4 quarts = 1 gal (US) = 3.875 L; 1L = 0.001 cubic meter
Label on a bottle of sparkling mineral water: 16.9 FL OZ (1.06 PT) (500 mL); L (liter) is the basic unit of volume in metric system, but technically not an SI unit. Liter is accepted for use in SI. It is further explained below.
$1L = 10^{-3}\ m^3 = (10^{-1}\ m)^3 = 1\ dm^3$. 1 liter = 1 cubic decimeter.

Typical problem on volume:

What is the volume of a container 50 cm by 40 cm by 100 mm?
It is noted that the prefixes of the three units are not identical. We will therefore, rewrite the units with uniform prefixes in decimeter (dm).
Solution:
50 cm x 40 cm x 100 mm = 5 dm x 4 dm x 1 dm = $\underline{20\ dm^3}$
Rounded off to 2 sf's.

Typical shapes and volumes:

Cone	$V = 1/3 \cdot \pi \cdot r^2 \cdot h$…(r=radius of circular base)
Sphere	$V = 4/3 \cdot \pi \cdot r^3$……(r = radius of sphere)
Pyramid	$V = 1/3 \cdot B \cdot h$……(B=base area; h=height)

Great Pyramid of Giza, Egypt:

Volume = $1/3 \cdot 137$ m high $\cdot$ 55, 000 m^2 footprint area = 2.51 Mm^3
Mass = 5.9 Mt, (1 t is metric ton = 1 Mg; 1 Mt = 10^6 t)

Density =5.9 Mt/2.51 Mm^3 = 2.35 t/m^3 = 2.35 Mg/m^3
For more data on earth's density, refer to p.31

10. SI UNITS IN PHYSIOLOGICAL MEASUREMENT

10.1 Radiation measurement

TABLE-14: Conversion Factors- radiation units

Quantity	SI Unit	Common Unit
Radioactivity	**becquerel (Bq)**	curie (Ci)
Absorbed Dose	**gray (Gy)**	rad
Dose Equivalent	**sievert (Sv)**	rem
Exposure	**coulomb/kilogram (C/kg)**	roentgen (R)

Common Units:

1 curie	=	3.7 x 10^{10} disintegration per second
1 becquerel (1/s)	=	1 disintegration per second
1 millicurie (mCi)	=	37 megabecquerels (MBq)
1 rad	=	0.01 gray (Gy)
1 rem	=	0.01 sievert (Sv)
1 roentgen (R)	=	0.000258 coulomb/kilogram (C/kg)
1 megabecquerel (MBq)	=	0.027 millicuries (mCi)
1 gray (Gy)	=	100 rad
1 sievert (Sv)	=	100 rem
1 coulomb/ kilogram (C/kg)	=	3 880 roentgens

Absorbed dose:
1 rad = 0.01 J/ kg = 0.01 Gy

Dose equivalent:
1 rem = 0.01 J/ kg = 0.01 Sv

Radio-activity:

curie (Ci)	becquerel (Bq) (SI)	dis/ min
1	**3.7 x 10^{10}**	2.22 x 10^{12}
2.7 10^{-11}	**1**	60
4.5 10^{-3}	**0.017**	1

Exposure:

R	C/ kg (SI)	MeV/cm³ (air STP)
1	**2.58 x 10^{-4}**	6.77 x 10^{4}
3.876 x 10^{3}	**1**	2.628 x 10^{8}
1.477 x 10^{5}	**3.811 x 10^{-9}**	1

1 milliroentgen mR = 0.258 microcoulombs /kilogram
Mev = million electron volt = a non-SI unit of energy
Coulomb/kg (C/kg)

Activity Release Rate:
1 Ci /d = 3.7 x 10^{10} Bq/d

= 4.287 x 10^{5} Bq/s (SI)

10.2 Light measurement

Measuring light is important when designing a room's lighting or preparing for a photograph. As the name suggests, over-illumination is basically too much of a good thing: excessive light for the space, task, or activity. This can mean lighting where it is not needed.

Let us be familiar with the following terms-

- Candela (cd) is the SI base unit of luminous intensity.
  The light from a lamp is measured in candela cd which is lumens per steradian = lm/sr.
- The lumen (lm) is the SI unit of luminous flux = 1/683 of a watt.
- Lux (lx) is the SI unit of illumination. The light that falls on a surface is expressed as illumination (lux = lm/m^2) which is the definition of lux= lumens per square meter.
- The human eye sees this visually in terms of brightness, or luminous flux that is measured in lumen lm= cd x sr.
- A photometer is a device that can measure the light intensity at different points of a light beam.

Sample problem

A 3 klm source spreads the light uniformly over 10 m^2 of wall, what is the mean illumination on the wall in lux?

Ans: 3 x 1000 lm/10 m^2 = **300 lx**

11. EXAMPLES OF CONVERTING TO SI UNITS

I. Conversion of traditional to SI derived units

Convert

i) 14.2 kg/cm^2 to N/m^2 (Pa)......... Pressure
ii) 115 lb/cub ft to kg/m^3.............. Density
iii) 22 Btu/(hr-ft-°F) to W/(m-K)...... Thermal conductivity
iv) 56 centi-poise to N $s\text{-}m^{-2}$........... Dyn. Viscosity
v) 61 Igpm to m^3/sec.................... Volumetric flow rate
vi) 11 kg/s water to kmol/s................. Flow rate
vii) 71.3 Ci/day to Bq/s.................. Radio-activity release

Solution

i) $14.2 \times 0.9807 \times 10^5\ N/m^2 = 1392594\ N/m^2$
= **1.39 MPa** [round off to 3 SF's]

ii) $115 \times 16.08\ kg/m^3 = 1849.2\ kg/m^3$
= **1.85 Mg/m^3** [round off to 3 SFs]

iii) 22×1.731 W/(m-K) = 38.082 W/(m-K)
= **38.1 W/(m-K)** [round off to 3 SF's]

iv) 56 centipoise = 56×10^{-2} poise x 0.1 $N\text{-}s/m^2$
= 0.056 N s $\text{-}m^{-2}$
= **56 mN s m^{-2}** [round off to 2 SF's]

Note: Be careful that m stands for milli in mN and meter in m^{-2}.

v) 61 Igpm = $61 \times 7.576 \times 10^{-5}\ m^3\ s^{-1}$
= 4.6213×10^{-3}=4.62 milli- $m^3\ s^{-1}$, or **4.62 L/s**[to 3 SF's]
Note: use of liter (symbol L) is a better choice than milli-m^3

vi) $11\ kg \cdot s^{-1} = (11\ kg \cdot s^{-1})\ /(18\ kg/kmol)$
= **0.61 kmol. s^{-1}**

vii) $71.3\ Ci\ d^{-1} = 71.3 \times 4.287 \times 10^5\ Bq\ s^{-1}$
$= 305.6631 \times 10^5\ Bq\ s^{-1}$
=**30.6 MBq/s** [round off to 3 SF's]

Note: More problems are solved in Appendix-2, Mass and Weight.

II. Conversion of formula/correlation to SI

Problem

Dry out heat flux refers to a critical condition in which the liquid film on a heated surface disappears due to excessive heat flux. There are correlations borne of experimental data.

The following Chalk River Canadian correlation was derived in FPS units
$\Phi_o = 0.8126\ G_0^{0.7325} - 0.0006327\ G_0^{1.128}\ hc_o$, where

Φ_o = critical heat flux, 10^6 Btu ft^{-2} h^{-1}	$\mathbf{\Phi_1}$ = 10^6 W m^{-2}
G_o = mass flux, 10^6 lb ft^{-2} h^{-1}	$\mathbf{G_1}$ = kg m^{-2} h^{-1}
hc_o = critical coolant enthalpy, Btu lb^{-1}	$\mathbf{hc_1}$ = kJ kg^{-1}

note: subscript o stands for FPS units; subscript 1 for SI units

<u>Solution</u>

For Conversion factors, refer to chapter 10

$$\mathbf{3.154 \times \Phi_1} = 0.8126 \times (4.8868\ G_1)^{0.7325} - 0.0006327 \times (4.8868\ G_1)^{1.1128} \times 2.326\ hc_1$$

$$\mathbf{\Phi_1 = 0.02045\ G_1^{0.7325} - 0.0163744\ G_1^{1.128}\ hc_1} \quad \mathbf{.....} \quad \mathbf{Ans.}$$

12.LIST OF APPENDICES

Useful notes:

- Treat each Appendix as a new hobby—one that brings measurable benefits to the body, mind, and spirit.
- As a yoga enthusiast, the author observes that performing exercises with spiritual awareness can raise overall well-being by about 8% and reduce stress and anxiety by nearly 10%. Long-term gains extend to cardiovascular health, immunity, musculoskeletal strength, mood stability, and protection against dementia.
- Time is the first and most fundamental SI base unit (see p.3). The message is simple: value time and do not waste it. Time ranks above wealth or material resources.

Using time wisely enables a conscious lifestyle, and letting go of the false ego is the true path to inner freedom.

APPENDIX 1 | SOME FUNDAMENTAL CONSTANTS IN SI SYSTEM

Out of a total of 26, the following eight Physical Constants are generally believed to be universal

Physical quantity	Symbol	Value in SI units
Avogadro constant	N_A	$6.022\ 141\ 29(27) \times 10^{23}\ mol^{-1}$
elementary charge	e	$1.602\ 176\ 565(35) \times 10^{-19}\ C$
Faraday constant	F	$9.648\ 533\ 65(21) \times 10^{4}\ C\ mol^{-1}$
gas constant [a]	R	$8.314\ 4621(75)\ J\ K^{-1}\ mol^{-1}$
magnetic constant [b]	μ_0	$4\pi \times 10^{-7}\ N\ A^{-2}$ (exact)
electric constant [c]	ϵ_0	$8.854\ 187\ 817... \times 10^{-12}\ C^2\ J^{-1}\ m^{-1}$ (exact)
speed of light in vacuum	c_0	$2.997\ 924\ 58 \times 10^{8}\ m\ s^{-1}$ (exact)
standard acceleration of free fall [d]	g_n	$9.806\ 65\ m\ s^{-2}$ (exact)

[a] or molar gas constant
[b] or permeability of vacuum
[c] or permittivity of vacuum
[d] or standard acceleration of gravity

APPENDIX 2 | MASS AND WEIGHT

A layman uses the terms mass and weight interchangeably in daily speech, because the value of $g = 9.81\ m/s^2$ is pretty much the same everywhere on earth. Weight scales in the stores show an estimate of mass, and both are expressed in kilograms or pounds.

In science and engineering, however, it is not acceptable, as explained in Fig.-1. Using lb, kg, or ton for mass as well as weight was one of the ills of the erstwhile customary CGS or FPS unit systems, which has been addressed by adopting SI. Now different values are used for mass and weight.

Mass (symbol, m) is defined as the quantity of matter contained in the object. Mass is an absolute property of the object that does not change from one location to another.
Kilogram (symbol, kg) is the base SI unit of mass.

Weight (symbol, F_w) on the other hand, is defined as the force of gravity of an object. Weight or any other force is a derived SI unit measured in newtons (symbol, N). Weight, being a derived unit in SI, is calculated from the equation $F_w = m \times g$. mass multiplied by acceleration due to gravity.

Fig 1a

Fig. 1b

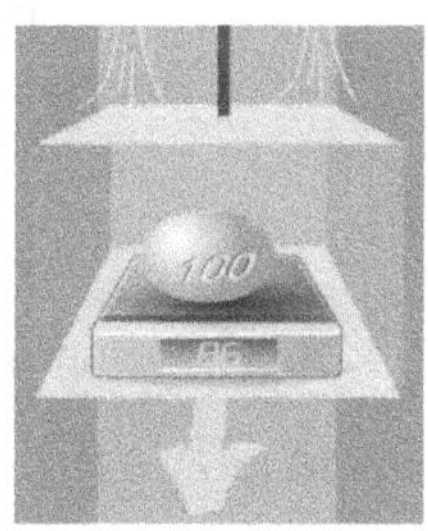

Fig. 1c

Refer Fig.-1: a 100 kg mass put on a measuring scale on the floor.

a) on earth, the scale reads 100 kgf, an estimate of mass. This shows on earth 100 kgm = 100 kgf. Multiply by 9.81 to convert to newtons = 981 N
b) on a space ship orbiting between stars, the scale reads 0 kgf or 0 N. Weightlessness.
c) in an elevator traveling downward at sufficient speed, the scale reads 86 kgf, a measure of apparent weight = 844 N

Note: The mass in each case is the same 100 kg

Effect of Gravitation:

Weight is often measured with a spring scale, and values are different if the spring is oriented vertically or horizontally due to gravity effect.

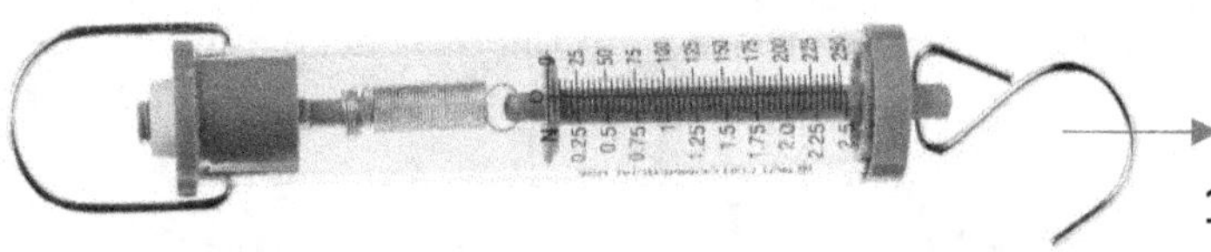

1 N

Horizontal orientation – gravity is not involved
(Same 1 N on earth as well as on moon)

0.16 N or 1 N

Vertical orientation – gravity is involved
(1 N on earth, or 0.16 N on moon)

Think Newtons:
In order to think directly in terms of force in newtons, following will help.

Weight of a medium size chocolate bar of mass 102 gram =
1.0006 N = 1 N *(1 significant figure)*

ii) Weight of a heavyweight wrestler of mass 102 kg =
1 000.6 N = 1 kN *(1 SF)*

iii) Weight of a large male elephant of mass 7 tons =
62 285N = 62.3 kN *(3 SF)*
(7 short tons=14 000 lbm, or 6 349 kg)

iv) Capacity of Forging Press of 15 000 tonnes =
147 MN *(2 SF)* (1 long ton, or tonne = 1 000 kg)

Problem on Energy:

Convert 400 kilowatt hours of electrical energy to SI unit.

Solution:

$400 \cdot 10^3 \cdot 60 \cdot 60 = 1.44 \cdot 10^9$ Watt-second = **1.4 GJ**

Problem on Mole:

What is the mass in kilograms of 10 kilo-mol of uranium atoms?

Solution:

1 mol of uranium = 238 gram = 0.238 kg

1 kmol = 1000·0.238 kg = 0.238 Mg

10 kmol = 2.38 Mg = **2.4 Mg** (2 SF)

APPENDIX 3.1 | SI - THE GROWING NEED OF MANKIND

- It was noted that the SI units were introduced in the 60s for scientific research, technological advancements in space and nuclear and various fields to facilitate precision, eliminate mishaps and increase reliability.
- Now applications of SI units can be found in home and office too during cooking, baking, home improvement, automotive and transportation.
- Some examples are illustrated

Cooking and baking

In New Jersey, U.S.A. my 14-y/o granddaughter prefers the measures of cups and ounces in her baking recipe.

For example:

Butter 1-1/2 c	sugar 1-1/4 c	Brown sugar ¾ c
large eggs 3	vanilla extract 3tsp	
All-purpose flour 2 c	Cream cheese 8 ounces	

Baking cocoa 1-1/2 tsp 12 other ingredients

If the food recipe is written in metric units, there is a conversion chart from metric ounces to cups and vice versa on the next page.

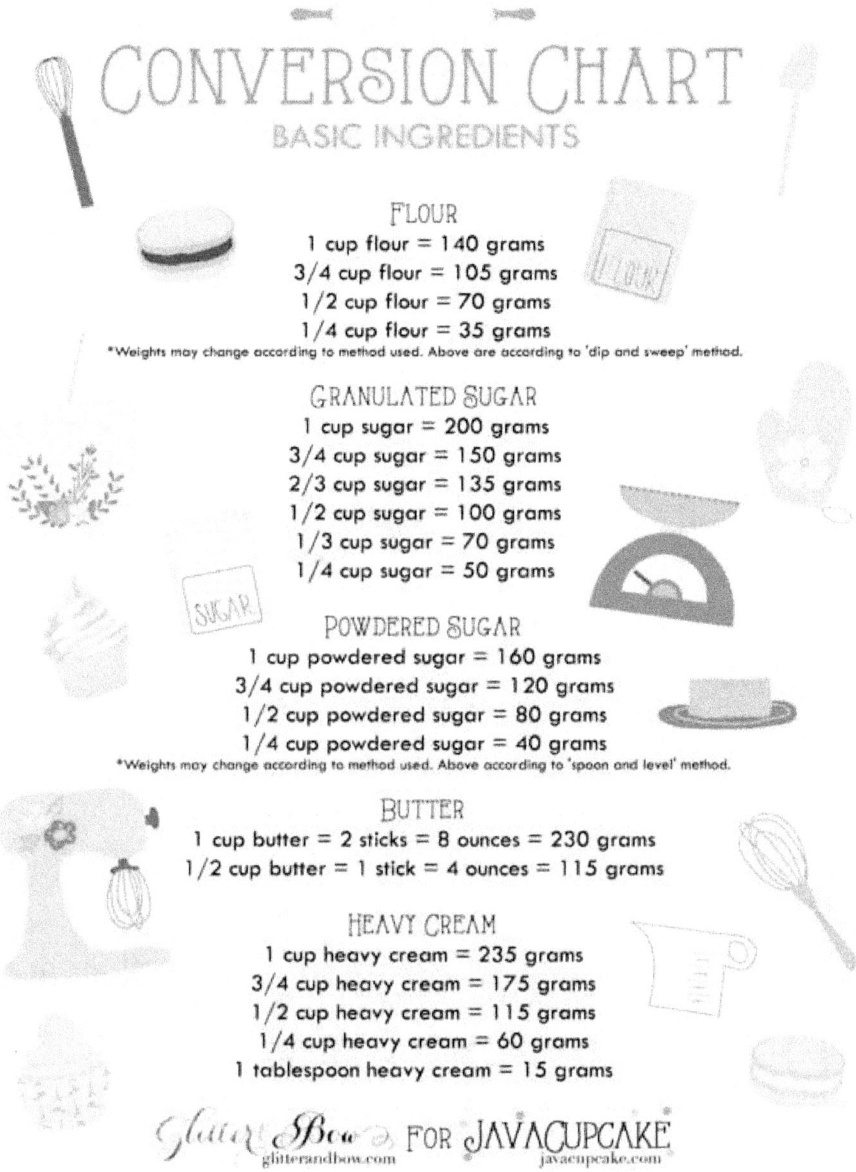

CONVERSION CHART

BASIC INGREDIENTS

FLOUR

1 cup flour = 140 grams
3/4 cup flour = 105 grams
1/2 cup flour = 70 grams
1/4 cup flour = 35 grams

*Weights may change according to method used. Above are according to 'dip and sweep' method.

GRANULATED SUGAR

1 cup sugar = 200 grams
3/4 cup sugar = 150 grams
2/3 cup sugar = 135 grams
1/2 cup sugar = 100 grams
1/3 cup sugar = 70 grams
1/4 cup sugar = 50 grams

POWDERED SUGAR

1 cup powdered sugar = 160 grams
3/4 cup powdered sugar = 120 grams
1/2 cup powdered sugar = 80 grams
1/4 cup powdered sugar = 40 grams

*Weights may change according to method used. Above according to 'spoon and level' method.

BUTTER

1 cup butter = 2 sticks = 8 ounces = 230 grams
1/2 cup butter = 1 stick = 4 ounces = 115 grams

HEAVY CREAM

1 cup heavy cream = 235 grams
3/4 cup heavy cream = 175 grams
1/2 cup heavy cream = 115 grams
1/4 cup heavy cream = 60 grams
1 tablespoon heavy cream = 15 grams

Glitter & Bow FOR JAVACUPCAKE
glitterandbow.com javacupcake.com

APPENDIX 3.2 | SI IN THE PHARMACEUTICAL INDUSTRY

SI units, such as milligrams, milliliters, and degrees Celsius, ensure standardized measurements for medications, fluids, and body temperature. This standardization minimizes the risk of medication errors, enhances patient care, and promotes consistent practices in healthcare settings.

METRIC MEASURES COMMONLY USED IN PHARMACY AND MEDICINE	
METRIC MEASURES OF LENGTH	
100 centimeters (cm) =	1 meter (m)
10 millimeters (mm) =	1 centimeter (cm)
1,000 micrometers (μm) =	1 millimeter (mm)
METRIC MEASURES OF VOLUME	
1,000 milliliters (mL) =	1 liter (L)
1,000 microliters (μL) =	1 milliliter (mL)
METRIC MEASURES OF WEIGHT	
1,000 grams (g) =	1 kilogram (kg)
1,000 milligrams (mg) =	1 gram (g)
1,000 micrograms (μg or mcg) =	1 milligram (mg)

Useful notes: Pharmaceutical industry has been consistently working to provide quality Rx, except there is need of affordable price.

The OTC over-the counter medicines are confusing due to lesser regulation, higher risk, lower transparency and public information.

Be aware of the benefits provided by your employer, particularly for anxiety and mental health. Practice meditation and deep breathing daily to feel better at home and work as well as to cut down insurance costs.

There is a joke- *When sick, visit the doctor, because doctor must live. Visit the pharmacy to buy the Rx, because the pharmacist must live. Do not eat the Rx because you must live. This means for illnesses-free living, practice yoga.*

APPENDIX 3.3 | NUTRITION LABEL

Regulations require that food and beverage manufacturers display nutrition facts label on their products, a sample is shown below.

Nutrition Facts

8 servings per container

Serving size	**2/3 cup (55g)**
Amount per serving	
Calories	**230**
	% Daily Value*
Total Fat 8g	**10%**
Saturated Fat 1g	**5%**
Trans Fat 0g	
Cholesterol 0mg	**0%**
Sodium 160mg	**7%**
Total Carbohydrate 37g	**13%**
Dietary Fiber 4g	**14%**
Total Sugars 12g	
Includes 10g Added Sugars	**20%**
Protein 3g	
Vitamin D 2mcg	10%
Calcium 260mg	20%
Iron 8mg	45%
Potassium 240mg	6%

* The % Daily Value (DV) tells you how much a nutrient in a serving of food contributes to a daily diet. 2,000 calories a day is used for general nutrition advice.

(For educational purposes only. These labels do not meet the labeling requirements described in 21 CFR 101.9.)

Serving size- suggested by the manufacturer
Calories- Calories in one serving, based on total 2,000 calories a day
Total fats- in three categories
- Saturated fats to account for no more than 5%-6% of calories intake
- Unsaturated fats- are as bad as saturated fats;
- Trans fats are worse. That explains why FDA banned these in 2018. Yet small quantities are found in some packaged foods. Be careful

Cholesterol – It is best to stay below 300 milligrams of cholesterol per day of the 2000 calorie diet
Sodium below 2,300 mg (600 mg for each of your three daily meals and less than 200 mg for two daily snacks.
Carbohydrates can be complicated, with a good side and a not-so-good side. The nutrition label offers a glimpse at both with information on dietary fiber (the good) and sugar (not-so-good).
Dietary fiber - Aim to get at least 25 to 35 grams of fiber every day.
Added sugars contribute to empty or non-nutritional calories that can lead to weight gain. Hence better to avoid, if you can.
Protein Aim for 20 to 30 grams of protein per meal.

APPENDIX 3.4 | VITAMINS AND MINERALS

Vitamins and minerals help your body grow, develop and work. Let us understand the following terms.

Percent daily value (% DV)
The righthand column of a nutrition facts label lists the percentage of the daily value (written as % DV) for each nutrient in one serving of a food or drink. A nutrient with a 5% DV or lower is considered to have a low daily value. A nutrient with a 20% DV or above has a high daily value.
Overall, it's generally better to have a:
Lower % DV for added sugar, sodium and saturated fat.
Higher % DV for fiber, vitamins and minerals.

When you browse the hundreds of vitamins and supplements at your local grocery or health food store, which one do you choose? The cheapest, right? Aren't all vitamins the same?

Not at all. Much like the rest of the groceries you buy, vitamin brands significantly vary in both quality and price. Quality of ingredients, bioavailability, manufacturing standards, and additives all affect your ability to benefit from your vitamins.

Choosing cheap supplements is a bit like opting for a $1.99 pack of frozen hot dogs, instead of cured and freshly sliced meat from the deli. The difference in quality is apparent from production to nutritional value.

Generally, it would appear the nutrition data is going from metric to SI.
However, most people are not aware of the math behind the label. For example, a medium size banana (125 grams) provides 112 calories, or 468 kJ. (Conversion: 1 Calorie or 1 kcal = 4.18 J)

APPENDIX 4 | SOME FAMOUS UNIT CONVERSION ERRORS

1. The Mars Climate Orbiter

Designed to orbit Mars as the first interplanetary weather satellite, the Mars Orbiter was lost in 1999 because the NASA team used metric units while a contractor used imperial. The $125m probe came too close to Mars as it tried to maneuver into orbit and is thought to have been destroyed by the planet's atmosphere. An investigation said the "root cause" of the loss was the "**failed translation of English FPS units into metric units**"

2. The Vasa warship

In 1628, crowds in Sweden watched in horror as a new warship Vasa sank less than a mile into her maiden voyage, with the death of 30 people on board. Armed with 64 bronze cannons, it was considered by some to be the most powerful warship in the world. Experts who have studied it since it was raised in 1961 say it is asymmetrical, being thicker on the port side than the starboard side. One reason for this could be that the **workmen were using different systems of measurement, on the two sides of external walls**. Archaeologists have found four rulers used by the workmen who built the ship. Two were calibrated in Swedish feet, which had 12 inches, while the other two measured Amsterdam feet, which had 11 inches.

3. The "Gimli Glider"

In 1983, an Air Canada flight ran out of fuel above Gimli, Manitoba. Canada switched to the metric system in 1970, and the plane is reported to have been Air Canada's first aircraft to use metric measurements. The plane's onboard fuel gauge was not working, so the crew used measuring "dipsticks" to check how much fuel the plane took on during refueling. Things went wrong when they converted this measurement of volume into one of weight. They got the number right, but the unit wrong - mistaking pounds of fuel for kilograms. As a result, the plane was carrying about half as much fuel as they thought. Luckily, the pilot was able to land the plane safely on the Gimli runway, giving the plane the nickname "Gimli Glider".

4. The Hubble Space Telescope

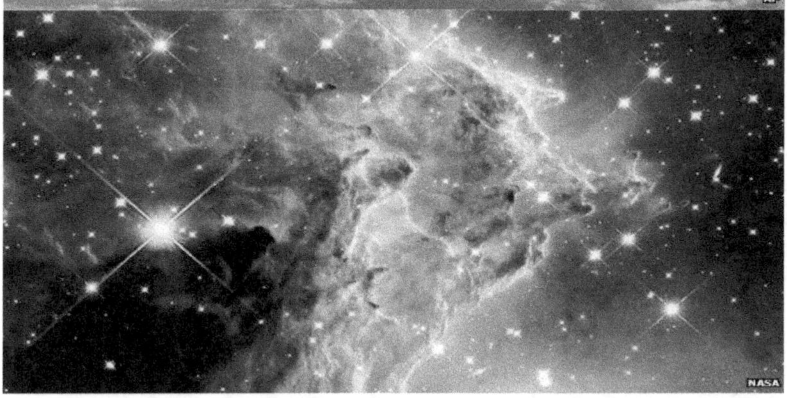

The Hubble telescope is famous for its beautiful space images and is considered a great success for NASA. However, it got off to a very rocky start. The first images sent back by the telescope were fuzzy because the telescope's main mirror was too flat. It wasn't out by much - only 2.2 microns, or about 1/50th the thickness of a human hair - but this was enough to put the project in jeopardy. One theory is that a speck of paint on a device used to test the mirror resulted in distorted measurements. Luckily, scientists managed to fix the problem in 1993, using an instrument called the Corrective Optics Space Telescope Axial Replacement (Costar). This cancelled out the error in the main mirror, by matching it in reverse.

5. Big Ben

The Big Ben bell at the Houses of Parliament in London cracked during testing in 1857 and was melted down to be recast. But the new bell, winched into position over three days in 1859, also quickly cracked. Disputes raged over who was at fault - there was even a libel case. One theory is that the massive hammer, at 6.5 hundred weights, was too heavy - at least for the alloy, the bell was made from (seven parts tin to 22 of copper). The foundries that cast the bells had always argued this material was too brittle. The second bell was not replaced (it is still cracked), just rotated by an eighth of a turn. The hammer, however, was replaced by a lighter one.

6. Christoffer Columbus discovered America?

Serious mathematical mistakes were made by Christoffer Columbus in calculating earth's size and shape, longitudes and distances in his 1492 first voyage from Spain to India. He realized that the New World of Indies he had reached in his westward voyage from Europe was not India, China or Indonesia. The new world became the United States of America.

7. **Stonehenge model**

In the 1984 mockumentary, this is a Spinal Tap, the members of a fictional rock group ordered a model of a Stonehenge megalith for their stage show - but the note written on a napkin mistakenly asks for a model 18 inches tall, instead of 18 feet. Curiously, and probably coincidentally, the British rock band Black Sabbath had experienced the opposite problem during its Born-Again tour in 1983. Its replica of Stonehenge was so big, that it got in the way of the band, and very few of the "stones" would fit on the stage. **One version of the story says there was a mix-up between meters and feet.**

APPENDIX 5 | SI UNITS IN NUCLEAR POWER

Brief history of nuclear power: EBR-1 (Experimental Breeder Reactor-1) in 1951 in the USA pioneered nuclear electricity development.
In 1954, the Obninsk APS-1 in the USSR was the first nuclear power plant to provide 5 MW of electricity to the grid.
Since then, significant progress has been made to generate electricity from the atomic fission. Nuclear power is not only the most powerful source, it is also most reliable because we have enough Uranium on the planet to generate energy for the next 70-80 years.

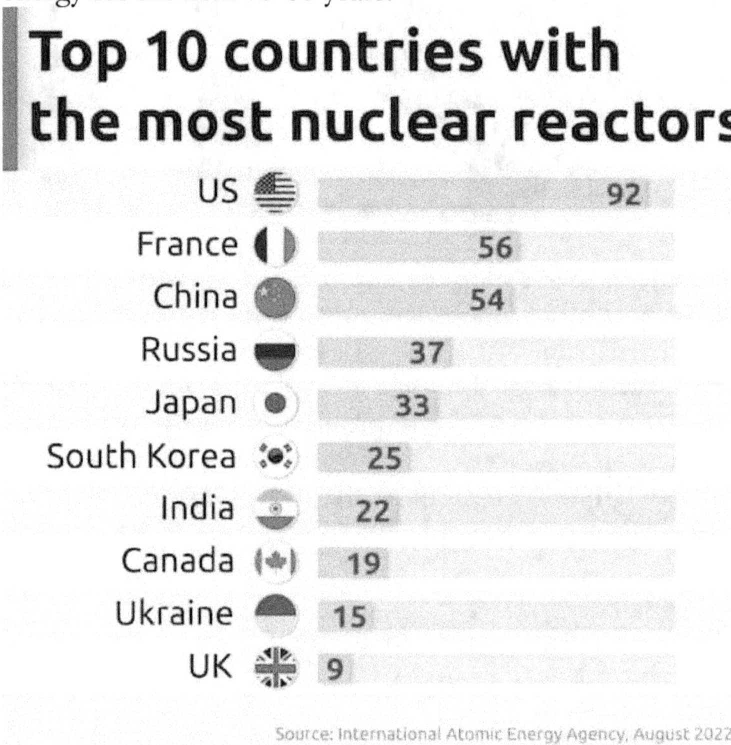

As of May 2024, there were 440 nuclear power reactors in operation in 32 countries around the world and generate about a tenth of the global electricity. The USA is the largest producer of nuclear power at 92 reactors while France has the largest share of electricity generated at 56 reactors.

Currently, there are 59 nuclear reactor units under construction worldwide. China ranked first with 25 units (17 365 MW), followed by India with seven units (6028 MW). Technical parameters for a Gen III+, the most popular LWR NPP- PWR are summarized below

Parameter	Value	**Unit**
Reactor thermal power	2998	**MW**
Pressurizer pressure	15.47	**MP_a**
Average temperature at core inlet	285.1	**°C**
Average temperature at core outlet	323.8	**°C**
Average coolant heating in core	38.7	**°C**
Total mass of water in primary circuit	198655.0	**kg**
Water mass in the pressurizer	18000.0	**kg**
Coolant flow through 1 MCP	4542.0	**$kg.s^{-1}$**
Total mass of water in secondary circuit	183768.0	**kg**
Pressure in the SG	6.136	**MPa**
Steam temperature in SG	277.0	**°C**
Steam mass flow from one SG	536.6	**$kg.s^{-1}$**
Feedwater temperature	216.9	**°C**

The public confidence in nuclear power is still shaken due to three major accidents in a short span of 32 years. 1979: Three Mile Island in the United States, 1986: Chernobyl, Russia, 2011: Fukushima, Japan

In each case, though the reactor type was different yet involved a similar scenario – fuel melting caused by mechanical failure and human error.

On the one hand, nuclear industry is yet to learn from past errors, being slow in incorporating lessons learned nor has created the safety culture required for sustainable nuclear power. There are growing concerns about nuclear disarmament, peace, conflict and waste management, non-proliferation, arms control, and the NPT.

Moreover, all the reactor vendors are slow to adopt SI system.

On the other hand, there is good news on the horizon!

APPENDIX-5.1 I TERRAPOWER MSFR

Good News: American philanthropist Bill Gates is spearheading an exciting new approach Natrium related to nuclear energy development in the future, in which units are SI, and fuel melting is avoided.

His company, Terra Power, recently broke ground on the first-ever Natrium plant in Kemmerer, Wyoming. This advanced nuclear facility aims to revolutionize how we generate power in the U.S. and around the world. The plant utilizes next-generation nuclear technology, relying on low-pressure sodium cooling instead of high-pressure water. By doing so, it enhances safety and simplifies the process. The plant is expected to come online by 2030, contributing to America's energy independence and the fight against climate change.

Terra Power's molten salt fast reactor major parameters:

-American patent design using SI units.

-345 MWe/840 MWt (Carbon-free electricity)

-Efficiency: lot more efficient than advanced GIII+ LWRs

-Uses: 50% less concrete, steel or skilled labor

-Low operating pressure

-Operating life up to 80 years

APPENDIX 6 | PROGRESS OF SI UNITS IN THE USA

The SI unit system has become the dominant measurement system in science, technology and international commerce. In recognition of this fact, the US Congress designated the metric system of measurement as the preferred system of weights and measures for trade and commerce. But it is not mandatory to use SI

Therefore, the USA stands out as the only industrialized country in the world that has yet to adopt the SI system legally for general use.

Engineers in the American manufacturing industry are comfortable in the conversion of SI specifications received from foreign clients while preparing drawings and documents for customer approval. It is hoped that an American people will think metric and not simply convert to metric. As seen in Appendix 4, conversion is confusing and liable to error.

In my 8 years of teaching at the Penn State University, the students were keen to learn using SI directly. The prescribed textbooks were using both- the traditional as well as the SI units.

Lately, in some States, road signs on highways are displayed in metric units.

APPENDIX 7.1 | PROGRESS OF SI UNITS IN INDIA

The author started his engineering career in 1955 studying for the bachelor's degree in mechanical engineering in Delhi, India. At that time the Imperial FPS units were common, metric CGS units were being introduced in schools and colleges, and SI was unknown.

His involvement spanning over 50 years has given firsthand experience with the traditional CGS, MKS and FPS units, the impracticality of which led to pioneer the use of SI in nuclear power plant technology and components manufacture indigenously.

In 1957, Government of India took a bold step to introduce a decimal currency scheme for common people where 100 paise formed the new rupee, versus 64 paise in the older system that was existing since the third century BCE.

Currently, majority of Indian people including street vendors prefer to use metric units in weights and measures and day-to-day commercial transactions with ease. You hardly meet people who talk pounds and ounces, feet and inches, bighas and acres in India. It is considered illegal to advertise for sale or purchase a material commodity in non-metric units.

In schools/colleges and technical institutions, the use of metric units as well as SI is becoming quite popular.

Thanks to yogic science in yesteryears, India was advanced in many branches of space science & nuclear technology, metallurgy, and mathematics, navy and shipbuilding, astronomy, and astrology.

A glimpse into the pre-historic units of time in India has been presented in the next Appendix 7.2

APPENDIX 7.2 | TIME RECKONING IN HINDU SCRIPTURES OF INDIA

Introduction

Hindu units of time in ancient Vedic years ranged from microseconds to trillions of years, including cycles of cosmic time that repeat general events in cosmology.

Lengths are given in divine years (a.k.a. celestial or Deva years), where a **divine** year **lasts** for **360 solar (human)** years.
The system incorporates four epochs or "Yugas": Krita Yuga, Treta Yuga, Dvapara Yuga, and Kali Yuga. This cyclical understanding of time is deeply ingrained in Hindu tradition and offers a distinct perspective on the unfolding of existence, even to a villager, or un-educated farmer.

Hindu concept of Time

The Hindu concept of time is encapsulated within the framework of the Vedic Time System. This is "kala," and doesn't adhere to a linear progression like an arrow moving solely from past to future. Rather, it presents a sophisticated rhythmic and universal order. This rhythm encompasses everything from the rapid vibrations of atoms to the grand expansion of the cosmos. It unfolds within the Earth's geological processes, the changing seasons, and the life cycles of creatures.

Large Units of time (billions and trillions of years)

A maha-yuga in Hinduism consists of four yugas:432 000 (Kaliyuga), 864 000(Dwapar yug), 1 296 000 (Treta yug), 1 728 000 (Satyuga)
One day of Brahma is of 1 000 (maha-yugas) = 4.32 billion human years.
In terms of human years

- Two *kalpa*s constitute a day and night of Brahma= 12 hour each
- 30 days = 1 month of Brahma (259 200 000 000 human years)
- 12 months = 1 year of Brahma (3 110 400 000 000 human years)
- 100 years are the lifespan of Brahma (311 040 000 000 000 human years ~ 311 trillion years). Currently, the creator Brahma is in its 51st year.

Smaller units of time (micro-seconds to kilo-seconds)

paramāṇu	base unit	≈ 26.3 μs (atom)
ahorātram (sidereal day)	30 *muhūrta*	≈ 86.4 ks (24 h)
masa (month)	30 *ahorātram*	≈ 2 592 ks
ritu (season)	2 *masa*	≈ 5 184 ks
ayana	3 *ritu*	≈ 15 552 ks (6 month)
year	2 ayana	≈ 31104 ks

Time is a great healer

All that is hostile and heavy in the world is overcome when we keep the innards human faculties- *antahkaran-* clean and purified. In this reckoning- passion, lust, money, and power are the commanding needs of a student and of a working householder. In the retirement phase of conscious withdrawal and reduced personal needs, the lifestyle is yoga-based.

Om is the primordial yogic sound. When properly uttered, it touches the ears, and the dormant inner spirit is awakened. Time then acquires a new dimension of bliss. Time no more hangs heavy, i.o.w., the unit system of time is seeking sweet happiness.

These esoteric time concepts of Hinduism are explained by Sankhya yog Appendix-7.3.

APPENDIX 7.3 | SANKHYA PHILOSOPHY OF INDIA

Hinduism owes a great deal to the Sankhya philosophy in the development of science and technology. Sankhya literally means enumeration.

Sāṅkhya analyzes the cosmos into two types of entities are *Prakṛti* or Nature and *puruṣa*-s or persons. Nature is singular, but persons are numerous.

- *Prakṛti*, or Nature, is comprised of three *gunas* or qualities. while in her manifest form she has twenty-three interdependent structures (*tattva*-s). The highest is buddhi, intellect. Total including purush and prakruti 25.

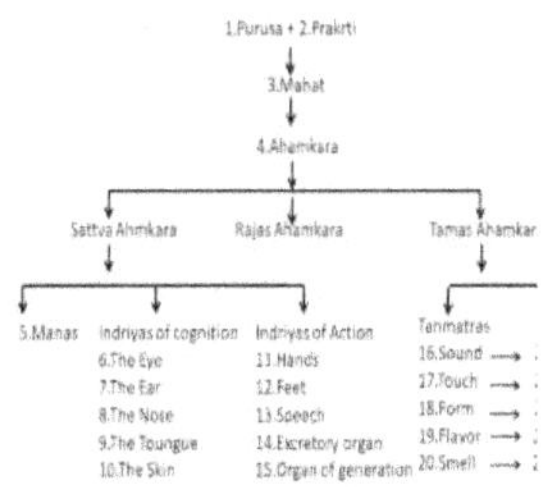

- *Buddhi*, discriminating, reasoning intellect (2)
- *Ahamkara* or ego-sense (3)
- *Manas*, the physical mind or brain (4)
- The five *gyanendriyas*, sense organs (9)
- The five *karmendriyas*, the organs of action (14)
- The five *tanmatras*, subtle elements or sense perceptions (19)
- The five *Mahabhutas*, gross elements

  earth, water, air, fire, and ether (24)

Not only the ancient Indians knew the system of numbers and zero, they had a broad vision on measures of metaphysics.

Count your Blessings One by One, and wonder what God has Done

13.REFERENCES

National Institute of Standards and technology, the International System of Units (SI), Publication NIST.SP.330-2019.pdf 2019

Brownridge, Dennis R. 'Metric in Minutes' Professional Publications, Ca 1994

Rustagi, R.S. 'A Simple Guide to SI units', Mumbai, India 1984

14. AUTHOR BIO AND ACKNOWLEDGMENTS

The author is a professional mechanical-cum-nuclear engineer who has promoted the peaceful use of nuclear power since 1959. He earned his B.E. in Mechanical Engineering from Delhi College of Engineering and pursued advanced studies at IIT Mumbai, JB Institute of Management Studies, and the Oak Ridge School of Reactor Technology in the United States. Over the years, he has taught college algebra, engineering mechanics, and SI units in both India and America.

His interest in SI units began in the 1970s at the Nuclear Power Corporation of India Ltd., during the development of India's indigenous Pressurized Heavy Water Reactors. At that time, NPCIL relied on a mix of FPS, CGS, and MKS units until the design team boldly adopted the new SI system. After moving to the United States in 1984, he worked in a nuclear-valve manufacturing company, using US customary FPS units for domestic clients and SI units for international customers. His enthusiasm for SI was renewed in 2010 while teaching at Penn State University–Schuylkill, where students preferred SI.

Prof. Rustagi retired in 2015; he now devotes his time to writing and practicing Yoga. His books—*The Metric World of SI Units* and *Yoga-on-Go*—reflect his belief that Yoga and SI are twin pillars of "unity in diversity."

He gratefully acknowledges the many technical professionals worldwide whose interactions inspired these works. Special thanks go to Rajat Rustagi for editorial support, to Saras Rustagi for his constant care and encouragement, and to Ms. Rea Rustagi, technical writer and mathematician from Columbia University, for her expert review. He also appreciates the contributions of Bharat Narang, Consulting Engineer (Nuclear Power) in Delhi, for editing and technical assistance in publishing the India edition.

On a personal note, he credits his yogic lifestyle for sustaining his mental clarity and recommends yoga as a foundation for any selfless and noble endeavor.

NOTES:

www.ingramcontent.com/pod-product-compliance
Lightning Source LLC
LaVergne TN
LVHW022025200926
848481LV00007B/2542

* 9 7 9 8 9 0 5 6 0 0 3 5 7 *